Past & Present

OXNARD

Opposite: Looking west, Park Plaza is pictured here around 1906, pre-pagoda (1910) and pre-Carnegie (1907). At far left is the steeple for the St. Paul's Methodist Episcopal Church at the corner of C and Sixth Streets. To the right is St. Joseph's Institute (1901) and the Santa Clara Church steeple (1904). Also seen in the image is the Oxnard Hotel (1899) before the roofline was replaced in 1910, and across Fifth Street, a dirt road at this time, is the Masonic Lodge (1901). (Courtesy of Jeff Maulhardt.)

PAST & PRESENT

OXNARD

Jeffrey Wayne Maulhardt

I dedicate this book to my three grandchildren, Reid Allan Hornbeck, Florence Elaine Hornbeck, and Evangeline Maulhardt Hopson. Grandpa's now pictures will be your then pictures. May you enjoy these pictures every now and then.

ISBN 978-1-4671-0726-6

Library of Congress Control Number: 2021938062

Published by Arcadia Publishing
Charleston, South Carolina

Printed in the United States of America

For all general information, please contact Arcadia Publishing:
Telephone 843-853-2070
Fax 843-853-0044
E-mail sales@arcadiapublishing.com
For customer service and orders:
Toll-Free 1-888-313-2665

Visit us on the Internet at www.arcadiapublishing.com

On the Front Cover: Plaza Park was the original gathering place. The townsite was laid out in January 1898 by James Driffill, the superintendent of the Pacific Beet Sugar Factory. By 1910, the city approved plans from Alfred F. Priest to build a structure over the water system, and he came up with a pagoda bandstand. Local contractor Thomas Carroll built the structure as well as dozens of prominent buildings in Oxnard. With time, restrooms were added then removed, a war memorial was added, and the corners of the park were rounded to accommodate traffic. (Past image, courtesy of Jeff Maulhardt; present image, courtesy of Aurelio Ocampo Jr.)

On the Back Cover: Pictured is the Louis Maulhardt bean threshing crew off Rice Road around 1910. Many of the early workers were of Portuguese descent from the Azores Islands and included the names Domingos, Dutra, Durate, Oliver, and Silva. Beans were harvested in the late summer or early fall depending on the weather. Crews would work in shifts around in the clock with a food wagon providing meals. The beans were piled up to be fed into the thresher, and the separated beans were sacked and sewn into 100-pound bags. (Courtesy of Jeff Maulhardt.)

Contents

Acknowledgments

Writing this book and collecting a new set of images during this COVID-19 pandemic time has been extremely challenging. With many places closed, boarded, and unavailable, I had to be extra creative and resourceful to come up with an acceptable book. Unfortunately, some images shared were not of print quality, while other sought-out images were unattainable. However, with the help of the following individuals, I think we have a pretty good document of some of the things past and present in Oxnard.

First, thanks to Caroline Anderson Vickerson of Arcadia Publishing for being open, responsive, and cooperative with me and my many excuses and delays. If I did not run out of excuses, I would still be delaying.

Thanks to my present-day photographer, Aurelio Ocampo Jr., from Red Sky Productions. Aurelio was prompt, open to coaching, and delivered winning images. These skills have been present since the days he was a part of the basketball championship team I coached at Driffill School. You deserve another trophy, Aurelio.

Also thank you to Gary Blum, who worked with Aurelio on the images for the Heritage Homes chapter.

A special thanks to Vince Behrens for not giving up on the image search and for being such a passionate Oxnard supporter.

Thanks to David Weigel for the Buddy Burger efforts and Steve Riley for the downtown Oxnard images.

Thank you to Deya Terrafranca and Elena Brokaw from the Museum of Ventura County for your efforts.

Acknowledgment goes out to our cat Smokey, who walked across my keyboard, laid on the scanner, and inspired me to take an occasional break in the action.

Finally, I appreciate my history mentors and supporters, Chuck Covarrubias, the late great Jim Gill, Frank Naumann, and newcomer Joe Pena.

Unless otherwise noted, all past and present-day images appear courtesy of the Jeff Maulhardt archives.

Introduction

It started with sugar. For 30 years, the farmers of the area experimented with various dry crops like barley and lima beans. They even tried growing potatoes, but by 1897, thanks to a commitment by the Oxnard brothers and their East Coast backers, a $2-million sugar factory was in the works. The site chosen was in the middle of a farming plain. To accommodate the workers of the factory, a townsite was laid out in January 1898 by James A. Driffill, the superintendent of the Pacific Beet Factory, soon to be renamed the American Sugar Factory in 1899 and finally the American Crystal Sugar Factory in 1934. Within months, whole buildings from Hueneme, Saticoy, and El Rio were transported across the flat land to their new home.

The factory brought workers to build, maintain, and run the facilities, and the new industry necessitated a larger workforce out in the fields. The first 30 years of dry farming barley and lima beans required little outside labor, and the population remained steady. However, the sugar factory brought in a whole new workforce of contracted laborers from Japan and Mexico. The Chinese railroad workers took on retail commerce. Oxnard's population became more and more diversified.

By 1910, Oxnard began to get the reputation as the biggest little city on the coast. The population was up to 2,000, and the town had a lot to offer. It had two large banks, the Bank of A. Levy and the First National Bank; the largest hotel between Los Angeles and San Francisco; a first-class library at the Carnegie building; an opera house; one of the largest sugar beet plants in the world, which employed up to 500 people during peak times; and beautifully built schools. By 1914, Oxnard had a state-of-the-art hospital in St. John's. By 1929, the first "talkie" theater in town was the Boulevard Theater (now Teatro).

After World War II, Oxnard took on many changes. By this time, the word was out. The population went from 8,519 in 1940 to 21,567. A decade later, the population almost doubled to 40,000. Many were employed by the Seabee Construction Battalion Center, the Naval Air Missile Test Center Base, or the West Coast Air Defense System Headquarters at Oxnard Airbase in Camarillo. By 1964, three military facilities employed 14,823.

Oxnard had plenty of room to grow, and as the city spread out, the farmland next to the city became more valuable. Farmers who struggled through the Depression years now found that their land near the city allowed them to cash out and try something else or in some cases find more affordable land in other parts of the state.

By the 1950s and 1960s, the original wood-structured businesses were torn down, and many of the brick buildings became unstable after the occasional earthquake reminded the building inspectors to improve the building codes. Roughly 500 buildings were demolished during the 1950s. Even the main building of the sugar factory was raised in July 1959.

In 1967, the city created the Redevelopment Agency, which targeted another round of building demolitions. Nearly all the buildings facing Plaza Park were removed and replaced. The city created an open-air pedestrian mall on A Street between Third and Sixth Streets, closing the street to automobiles and taking away the traditional cruise night. Many of the drive-in diners on the boulevard suffered.

Also hastening the economic decline of downtown was the development of the Esplanade Mall, which opened in 1970 at the north end of town.

Channel Island Harbor was dedicated in 1965, and soon after, two channels were dredged to accommodate the proposed boat docks. The channels would grow into many water fingers and become part of a larger dockside community. Hollywood by the Sea, Silver Strand, and Oxnard Shores continued to populate, and Oxnard focused on developing its coastline. The Esplanade Shopping Center was reimaged and rebuilt in 2000, and after the slow recovery from the recession of 2007, the Collection at River Park held its grand opening on November 15, 2012. The Junction at Wagon Wheel held its grand opening on January 23, 2019, and the next year, Amazon and the city announced that a giant fulfillment center would be built on a 75-acre parcel off Rice Road. With the city now full of residences and businesses, farming has become the past, and commercial retail has become the future.

CHAPTER 1

DOWNTOWN OXNARD

The Oxnard Hotel was built for $20,000 in 1898. On May 27, 1910, a fire ripped through the structure, causing extensive damage. Ernest Eastwood purchased the hotel and in September 1910 hired Abplanalp and Myers to renovate the hotel and give the building a new roofline, as seen is this picture taken a few years later.

The pagoda has been the central location for events, group photographs, music, dedications, and speeches. Early performances included the Oxnard Military Band, the St. Joseph's Boys Band, Evangeline Carroll, the Salvation Army Band, the Orquestra Mutualista Jazz Band led by Tony Avila, and the Mitchell Boys Choir. Over the years, many famous people have spoken at or stopped by the pagoda, including Edward Kennedy, Cesar Chavez, and George Lopez. The park has held the annual tamale festival and the weekly farmers market on Thursdays.

Mayor Manuel M. Lopez spearheaded the building of the War Memorial monuments at Plaza Park in 1991. The black granite panels were mounted on 15-foot-high pyramids forming a circular path. The monuments were designed by Elizabeth Nevandro-Perez. There are 132 names inscribed on the five monuments of those who sacrificed their lives in World War I, World War II, and the Korean, Vietnam, and Persian Gulf Wars. Among the strong advocates and committee members was Catherine Mervyn, a veteran of World War II.

The Oxnard Implement Company lined up its newly arrived Studebaker wagons. The company was started by Frank Borchard, A.D. Thatcher, Jay Spence, and E.C. Griffith in 1900. The wagons were used in the sugar beet fields. This image was taken in front of the open lot that became the Carnegie Library in 1907.

The east wing (left side) of the Carnegie building was added in 1923 and was designed by Alfred Priest and built by Thomas Carroll. Frank Burnham designed the original structure. The building also served as city hall in its early years. When a new library was built in 1963, the Carnegie became home to the chamber of commerce and later an art museum. In 1971, the Carnegie became the first building in Ventura County to be listed in the National Register of Historic Places.

PUBLIC LIBRARY. OXNARD. CALIFORNIA

The First Baptist Church began its first service in a converted train car in 1900. Two years later, the members built a 2,500-square-foot church. Two years after that, they had it moved to a lot on the corner of Fourth and C Streets, three years before the erection of the Carnegie Library. In 1926, the building was moved again, this time to the corner of Hayes and Second Streets in the Colonia area. Unfortunately, the steeple was removed and not replaced.

The Oxnard Hotel was opened on September 1, 1899, and when completed was the largest hotel between Los Angeles and San Francisco. The original roofline was replaced after a fire in 1910. Many famous people stayed at the hotel, including Rudolph Valentino. The last guest checked out on March 24, 1960. The building was replaced by the Guardian Savings & Loan. Today, the building hosts Western Dental & Orthodontics. (Past image, courtesy of Tom Laubacher.)

The Guardian Savings & Loan was organized in 1954 and was originally located at Sixth and B Streets. On June 30, 1963, a grand opening for the new building was held at the site of the recently demolished Oxnard Hotel at 455 South C Street. Another branch opened in the Esplanade area in 1970. By 1976, the company merged with Citizens Savings. Today, the building hosts Western Dental & Orthodontics. (Present image, courtesy of Aurelio Ocampo Jr.)

Pictured is Plaza Park looking west toward Fifth Street. The park has a long history of efforts to open up traffic through the middle it. In 1928, citizens took a vote on opening the middle of the park to traffic from Fifth Street. The division of the park failed by 28 votes of the 500 cast. By 1991, the city commissioned a plan to redesign traffic from Fifth Street around the park for two-way traffic along the south side of the park.

The Masonic Lodge was at the corner of C and Fifth Streets. The building was completed in 1901. By 1915, the lodge began running silent movies under the name New Victory Theater. In March 18, 1957, an earthquake caused enough damage to the building to require major repairs and finally demolition.

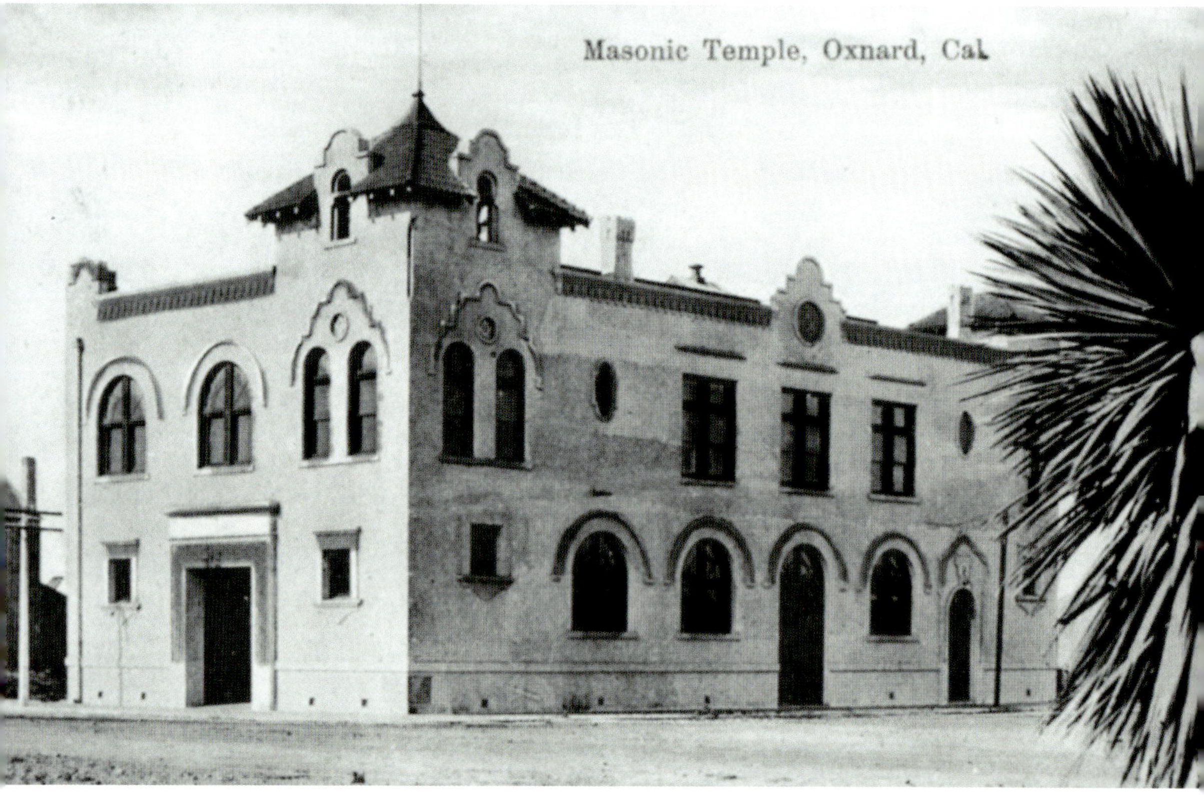

Masonic Temple, Oxnard, Cal.

The building at 511 C Street was constructed in 1927 in the Spanish Colonial style. In 1948, Dr. Nobel Powell purchased it. The building at 519 C Street was sold in 1955 to local educator and historian Madeline Miedema and was rebuilt by Adolph Schroeder, who also constructed the building at 521 C Street in 1947 for Dr. J. Robinson. The middle building was once home to the Oxnard Opera House, which Miedema wrote about for the *Ventura County Quarterly*.

Alvin Behrens began working at American Cleaners when he was 12 years old. By July 1956, Behrens and Carlo Silvio purchased the three dry cleaners owned by Max Koren: American Cleaners, the Society Drive-in Cleaners, and Band Box Cleaners. American Cleaners began operating in 1902 as Diefenbach's Cleaners. Behrens was also instrumental in reopening A Street between Third and Sixth Streets. Today, Behrens's son Vince continues the family business and downtown promotion. (Past image, courtesy of Vince Behrens.)

Toraichi and Shina Otani opened a grocery store on Oxnard Boulevard in 1908. After a stay at the internment camp at Gila River, they reopened their old building as a malt shop. By 1952, Isuto "Izzy" and Helen Otani opened Izzy Otani Fish Market at 608 A Street. By 1972, the shop expanded into a popular restaurant run by Steve Otani, who in turn handed the business to his children Karen and Genji. Karen's daughter Tani represents the fifth generation of Otani leadership. (Past image, courtesy of Karen Otani Baldonado.)

The building next to Otani's Seafood on the corner A and Sixth Streets was constructed for Crescent Garage. As early as 1915, Crescent was selling Chalmers, Hupmobile, and Stutz autos. By the 1950s, the building served as a pet store, Barney's Pet Barn. Today, the site is used as a parking lot for Otani's Seafood.

The Japanese Methodist Episcopal Mission Church was organized by Kasaburo Baba, a labor contractor who also served as a minister for the church. The building was dedicated in 1908. The steeple was added in 1910, and in 1940, the building was moved to the rear of the lot, closer to A Street. The congregation moved to a new location in 1965, and this lot became home to the New Hope Baptist Church and most recently the Church of Christ.

In May 1902, the corner of Fifth and B Streets became the home of Bank of A. Levy, founded by a Jewish immigrant from France, Achille Levy. The bank moved one block away in 1926. The building next became home to Poggi's Drug Store until 1942, when the Cooluris brothers opened a liquor store and delicatessen. The building was demolished in 1963. The Plaza 14 Cinemas opened in August 2005.

Achille Levy cofounded the Bank of Hueneme in 1889. By 1900, he opened the Bank of A. Levy in Oxnard. In 1927, his son Joe Levy oversaw the completion of the new bank at the corner of Fifth and A Streets that incorporated Greek features like high windows, rounded arches, and Doric columns. By 2014, the building was purchased and occupied by Barkley Insurance & Risk Management.

The Security First National Bank, at the corner of Fifth and A Streets, started out as the First National Bank before a merger in 1922. The building was designed by Los Angeles architect A.C. Martin. Among the original directors were Charles Donlon, James A. Driffill, Frank Petit, T.A. Rice, and Mark McLoughlin. Today, the corner building is occupied by the Mixteco Indigena Community Organizing Project. (Past image, courtesy of Steve Riley.)

Oxnard Theater was designed by Alfred Priest and built by Adolph Schroeder for J. Roy Williams at a cost of $100,000. It was completed by June 1928, on the eve of both the Great Depression and the birth of "talkies." Unfortunately for posterity, the Oxnard Development Agency turned the elegant theater into a temporary parking lot for the development of the short-lived A Street Mall.

The north side of Fifth Street between A and B Streets included Lynn Shoes, Harvard Hotel, Rodaway Shop for Men, and Marvin Drugs Store. They were all torn down, making way for Centennial Plaza, which includes Fresh and Fabulous, Sushi Way, Plaza Stadium Cinemas 14, Cold Stone Creamery, La Vero's Mexican Seafood, Subway, and Starbucks. (Past image, courtesy of Steve Riley.)

Rain's Shoes, at 428 South A Street, was situated next to the Charlotte Shop, Oxnard Department Store, and Kane's Men's Ware. Rain's gave way to Nate and Bernie's Kosher Style Delicatessen, and in 1971, Bill and Gloria Stuart opened BG's Café. By 2004, the Stuarts retired and sold to Jose and Veronica Rodriguez, who carried on the tradition. (Past image, courtesy of Steve Riley.)

The Safeway Store opened in 1927 at 401 South A Street. By 1950, the store relocated to 939 South Oxnard Boulevard, and the building was redesigned for the Woolworth store. By 1997, the business closed, and soon after, the building was purchased by David Feigin. He created a Woolworth museum in a portion of it, while the restaurant that occupied the front portion was Fresh and Fabulous. The museum recently closed in 2018, and Fresh and Fabulous relocated to Fifth Street.

The post office was constructed in 1939 by the Works Progress Administration (WPA), an agency set up during the Depression and part of Pres. Franklin Roosevelt's New Deal program. Built in the Italian Renaissance Revival style, the building features a panoramic mural in the lobby. The Treasury Department's Section of Painting and Sculpture commissioned Daniel Marcus Medelwitz to paint a mural in 1941. He was a Stanford professor and author of *Guide to Drawing*. The mural is entitled *Oxnard Panorama*.

Oxnard Grammar School was located at C and Third Streets and completed in 1901. The building burned down in 1923, and the district constructed Roosevelt School and, a few blocks away, Wilson School. By 1945, the Roosevelt School building was purchased by the city and converted into a civic center. In 1965, Viola Construction built a 17,534-square-foot replacement structure at a cost of $496,500.

The A&P Food Store (the Great Atlantic & Pacific Tea Company) started its Oxnard business at 137 West Fifth Street in the 1930s. By the 1940s, it moved to 235 West Fourth Street. In 1958, the store added a Van de Kamp Bakery. Today, the site is a two-story parking garage. (Past image, courtesy of Steve Riley.)

China Alley was an area between Seventh and Eighth Streets and Oxnard Boulevard and A Street. Nightlife came alive with gambling, opium dens, and prostitution. Many of the shops had false fronts, sliding doors, and a series of bells, buzzers, and secret doors leading to hidden stairways and underground tunnels. After a series of raids in the 1920s and a receding population in the 1930s, the alley's illegal activities became less frequent. Today, many respectable businesses like Tomas Café have taken root in the area.

Mel's Drive-In Diner was opened in the 1958 by Mel Buttram. This diner replaced Buttram's previous restaurant, Round Burger. Mel's was a popular Friday night spot after football games as well as on the A Street cruise nights. By 1976, the building was converted to a Sizzler Restaurant, and in 1997, the structure became home to Peter Spink, who opened Cabo Seafood Grill and Cantina.

The Flamenco Club, at 663 South Oxnard Boulevard, was started by John Thomas in the 1950s and offered a variety of entertainment, from comedians out of Las Vegas to live music. He sold to Pasqual Bravo in 1964. Today, it is home to the Tapatio Night Club. (Past image, courtesy of Steve Riley.)

The original Asahi Market was opened in 1907. By 1957, a new building was erected at 660 Oxnard Boulevard. The Asahi family was relocated along with nearly 700 Japanese during World War II. By 1950, there were 362 returning Japanese. Many of these resilient people became leading Oxnard citizens, including Nao Takasugi, who became the mayor from 1982 to 1991. Tsujio Kato became a dentist and also served as an Oxnard mayor from 1976 to 1982.

In 1908, Torachi Otani established a poolroom at 620 Saviers Road (Oxnard Boulevard). The business evolved into the Otani Bros. Malt Shop and in 1953 became a fountain and grill. The business was sold in 1956. Next door, Boulevard Theater opened in 1929 as the first "talkie" theater in the county. The theater began showing Spanish-language films in 1957 and later became the Teatro. It later served as a hidden recording studio for Bob Dylan, Willie Nelson, and Neil Young.

The building at 719 Oxnard Boulevard has gone through several businesses since 1928. The original business was the New China Café, preceding the Golden Chicken by one year. The advertisement in *Oxnard Courier* announced, "New building has new equipment and ready to serve you—Chop Suey and Noodles." By 1955, Cardona's offered radio and television repair and specialized in Latin American records. Today, the building is occupied by Rebecca's Juice Bar Y Cenaduria.

The Chinese arrived with the creation of the sugar factory, which was completed in 1898. In 1904, they built a hall for the Chinese Free Masons, Bing Tong Kong at 740 A Street. Here, they organized a fire department, a court system, and a place for funeral arrangements. In 1921, they moved the building to 743 South Oxnard Boulevard. The building was condemned in 1954, and they erected another structure at 749 South Oxnard Boulevard.

The Blue Onion Drive-In, at 733 South Oxnard Boulevard, opened in September 1956. In addition to indoor dining, there was drive-in service for 30 autos. The building was designed by Winton Hanson and was part of a chain of five restaurants from Goleta to Las Vegas. "We thought we finally made it," is what longtime resident Chuck Covarrubias remembered about the new dining addition to the city. Today, the building is occupied by El Chilito, serving Mexican food.

Buddy Burger, located at 600 Oxnard Boulevard, began as Wimpy's Hot Dogs before evolving into Wimpy's Hamburgers. In 1951, the restaurant was run by Wimpy and Ola Sheldon. By the 1971, Ernest and Ethel Wood took over. In 1978, a truck crashed into the building, and the restaurant reopened with a new structure and a new name: Buddy Burger. Chris Weigel stepped in for her mother, Ethel, in 1996 and kept the famous chili recipe alive. She sold the business and the chili recipe in 2012 to Jesse Bustillos. (Past image, courtesy of David Weigel.)

Dominick and Sandra DeMatteis opened Dominick's Italian Restaurant in 1940 at 477 North Oxnard Boulevard. The building was originally used as an auto sales and repair shop by Auto Electric Garage. Florence Henggler, Dominick's niece, and her husband, Bill, left New York for Oxnard in 1958 and took over the restaurant. Florence and Bill's youngest sons, Brian and Sean, eventually took over the business. The restaurant enters its seventh decade serving Oxnard's favorite Italian entrees. (Past image, courtesy of Brian Henggler.)

The L.G. Maulhardt Equipment Company opened at 350 South Oxnard Boulevard in 1939 after purchasing the International Harvester dealership from Joe Schreiner. Ten years later, a new building was designed by Walter D. Hessert and constructed by Bergseid Construction at 815 Oxnard Boulevard. It included 30,000 square feet of floor space and a 5,000-square-foot mezzanine. By 1970, Tom Coward Ford took over the building, and today, the site is home to the Super Thrift Store.

Art's Drive-In preceded the Colonia House. In 1941, Martin "Bud" Smith traded a couple of his coin-operated machines for the struggling restaurant. While he was serving in the Air Corps during World War II, Smith's mother, wife, and sister took over the 22-stool hamburger stand and transformed it into a thriving business. Smith grew the business into seven private dining rooms and two bars. He sold in 1967, and by 1988, the restaurant was closed. The site became a mixed-use residential and commercial set of buildings. (Past image, courtesy of Frank "Robey" Naumann.)

The barker, or "beckoning roadside chef," on Oxnard Boulevard waving in customers to the Colonial House across the street was to represent the hospitality of the restaurant. However, the NAACP did not see it the same way, and the chef was put back in the kitchen. While the orchards behind the chef have been replaced with houses, the 1948 Oxnard Lemon Company building, which is out of frame to the right of the chef, remains and has joined the Lemoneira family.

Aswell Trophy & Engraving was started by Jed Ashton out of the back room of his Foster Freeze building in 1965. By 1970, the business moved across the street to 134 Palm Drive. Finally, in 2000, Aswell Trophy moved into the old Al Schreiner building at 235 North Oxnard Boulevard. Today, the business is run by his son Trevor Ashton. (Past image, courtesy of Trevor Ashton.)

Foster Freeze, at 310 North A Street, was another popular A Street cruise night destination in the late 1950s and early 1960s. Owner Jed Ashton also opened Aswell Trophies across the street in 1965. The shop relocated to 235 North Oxnard Boulevard and continues as a family-owned business. Foster Freeze now operates as Win's Drive-In and carries on the tradition of Foster's famous corn burritos. (Past image, courtesy of Trevor Ashton.)

CHAPTER 2

AROUND OXNARD

On February 10, 1898, the *Ventura Signal* reported on the barbecue and dance at the "Colonia factory." Over 4,000 people attended, and Charles Donlon served as the floor manager for the dance contest. First prize went Jess Bates, who won a $5 fan. Music was supplied by the Ventura City Band. Also in attendance was Robert Oxnard, who was in town to meet with Supt. James A. Driffill and the construction engineers. The factory was completed by August 18, 1898.

The vineyard at the Oxnard Historic Farm Park was planted by Paul Belgum and Jeff Maulhardt with vines from Santa Cruz Island that date back to the 1880s. The vintage zinfandel variety compliments the 1870s winery built by Gottfried Maulhardt. Currently, the vineyards are maintained by Jim Thompson, David Young, and Terry Ball from the Master Gardeners of Ventura County.

The Santa Clara Chapel was built in 1876 and was originally located near the 101 Freeway and the Vineyard Avenue overpass. With the widening of the freeway in 1953, the church was moved to its current location near the Rose Avenue overpass. (Past image, courtesy of Mary "Ginty" Sargent.)

The Oxnard brothers built the Pacific Beet Sugar Factory in 1898. The following year, the company became the American Beet Sugar Factory, and in 1934, it was renamed the American Crystal Sugar Company. The Oxnard family has returned many times. Robert Oxnard lived in San Francisco and made biweekly visits. Benjamin Oxnard was welcomed to the city's 80th anniversary celebration. James and Thornton Oxnard have also visited. Thomas Oxnard attended several centennial activities in 2003, and in 2016, Bob Oxnard toured the remaining factory buildings.

Pictured is the interior of one of the two remaining warehouse buildings of the sugar factory over 100 years apart. While sugar is no longer being grown and processed in the area, berries are, including raspberries. Raspberries ranked fourth in agriculture product sales in 2019 behind strawberries, celery, and lemons. Today, one of the buildings is home to Maxco Supply, a company specializing in agricultural packaging.

For the first 30 years of dry farming barley and later lima beans, a small workforce was needed to plant and maintain the fields. With the introduction of labor-intensive crops like sugar beets, a larger workforce was required. The Japanese were among the first contracted field workers, with over 1,000 young single men recruited from Kumamoto Prefecture, Japan. By 1909, the Japanese accounted for 42 percent of the agriculture workforce. It was not until 1920 that half of the field labor came from Mexico.

The agricultural landscape of the Oxnard Plain has changed many times over in the 150 years of commercial farming. The first farmers encountered the wild mustard plant before the land gave way to the dry-farming crops of barley and then lima beans. Sugar beets, seen above in a field, were introduced in 1896, and two years later, it was the main cash crop. Citrus, row crops, and strawberries followed. In recent years, hoop crops like raspberries have produced over 64,000 tons of fruit a year. (Present image, courtesy of McGrath Family Farm.)

Lima beans arrived just north of Oxnard in Carpentaria in the 1870s. By the 1880s, Joseph and William Lewis left Carpenteria and began planting acreage around the county. The foggy coastal area of Ventura County proved fruitful for the dry bean. By 1903, the Ventura Implement Company began manufacturing threshers that were sold throughout the county. Above are piles of bean straw from threshing lima beans at the Bill Lenox Ranch around 1960. More recently, Lenox switched to the more profitable berry crops.

A large red barn was part of the Bob Jones ranch. His father, Roger Jones, was the foreman of the 320-acre Hobson Ranch in the 1920s. The five-story, 300-by-40-foot barn was used to store up to 1,300 tons of hay. Bob purchased the ranch and became one of the first farmers to grow strawberries on a large scale. His son Dick Jones continues farming. However, the barn burned down in January 2000 due to an electrical short.

Bell Ranch was a reference to the original landowner, Thomas Bell, who owned a 100-acre ranch off Rice Road and the current Camino Del Sol. Bell planted one of the first lemon orchards in the 1910s. His heirs sold the ranch in 1926 to Louis Maulhardt, and by 1953, his son Robert Maulhardt took over the ranch. After the property was sold and rezoned into commercial, Lombard Street was created. It runs almost approximately in front of the ranch home and is now home to Kanaloa Seafood.

Another image from the Bell Ranch shows the grandchildren of Robert Maulhardt on a Farmall tractor getting ready to work the lemon orchards. Directly behind the Maulhardt grandkids is where the present-day Deardorff Family Farms packing and cooling facility has been built. W.H. Deardorff started the sales and marketing company in Los Angeles in 1937 to consolidate crops from various growers. By 2012, the family built a 110,000-square-foot, state-of-the-art facility at 400 North Lombard Street.

The Samuel Naumann family arrived in Oxnard in 1898 with the building of the sugar factory. Naumann purchased his first 159 acres in the Ocean View District off Etting Road in 1901. Naumann's grandson Robert purchased a 3.5-acre parcel next to the Masonic Cemetery off Hueneme Road in the 1940s. Robert's son Frank R. Naumann took over and planted an avocado orchard, farming until the property was sold in 2014. Today, the property is home to the Coastal Village Apartments.

L.J. Rose Jr.'s ranch home, pictured around 1893, was christened Roseland. The Colonial-style home featured five bedrooms, a large glass-enclosed dining room, and a veranda that stretched around two sides of the house. The property also featured tennis courts and a one-room schoolroom. The property was sold to the McGrath family in 1909. John McGrath and his caregiver, Douglas Carty, were the last to live on the ranch. The house was moved to Camarillo Heights to make way for the Rose Ranch shopping center.

The south side of Fifth Street, looking east between B and A Streets in the 1950s, included many businesses, among them Boo's Stationery, Dolly Brigham's, Peacock's Ice Cream parlor, Saake's Jewelers, U&I Café, Thrifty's, and Security First National Bank. Today, the Church for the Nations takes up several lots and is bordered by an alley. The block ends with the former bank site, now home to the Mixteco Indigena Community Organizing Project. (Past image, courtesy of Steve Riley.)

In 1952, a committee led by John Maulhardt and Drs. Cloyce Huff, Charles M. Hair, and Nobel Powell raised the funds for an additional 75 beds for St John's Hospital. By 1992, the hospital relocated to 1600 Rose Avenue, and for several years, the abandoned building was subject to many ghost stories. Sycamore Senior Housing grew out of the old St John's Hospital and opened its doors in 2009. It is a low-income, tax-credit-assisted living facility for seniors and offers 228 units.

Albert Pfeiler displays his fleet of Overland autos around 1910 along Fifth Street looking toward Sixth Street. Like most farmers, Pfeiler was looking for the next cash crop to expand his harvest, in this case as an agent for the Overland Automobile Company. Overland was one of the first automobile companies and second behind Ford and the Model T. Today, the majority of autos are sold near the 101 Freeway along Auto Center Drive.

Beardsley & Son began operations in Oxnard in 1935 by selling custom fertilizers to local farmers. Robert Beardsley Jr. farmed several ranches and several hundred acres from Camarillo and present-day Beardsley Road as far west as Wooley Road in Oxnard. With time, his sons Robert III and later Tom would join the fertilizer business. Tom's daughter Melinda eased her way into the company by 2008 before taking over in 2017.

The Wagon Wheel Coffee Shop was part of a 64-acre Western-themed collection of shops, restaurants, and motels developed by Martin "Bud" Smith in 1947. After attempts to save the historical buildings fell short, demolition began in 2011. By 2018, signs of rebirth began for the 1,500 apartments, condominiums, and townhomes. Included in the development are restaurants, parks, a bowling alley, and businesses. The new Wagon Wheel community includes the Oxnard Flats, Park Place, Mayfair, and the Junction at Wagon Wheel.

Wagon Wheel Junction featured a bowling alley, shopping center, roller rink, and several restaurants, including the Wagon Wheel Steakhouse, El Ranchito, and the Trade Winds Polynesian restaurant. Rooms for the motel were $3 for a single and $4 for a double. Some of the street names from the original Wagon Wheel area have been reincorporated, including Buckaroo Place, Spur Drive, Winchester Drive, and Wagon Wheel Road. The present image shows a view toward Wagon Wheel Drive. (Past image, courtesy Museum of Ventura County Research Library and Archives.)

Sears served as an anchor building for the newly constructed Esplanade Mall, completed in 1970 and the county's first fully enclosed shopping center. The May Company anchored the opposite end of the 45-acre lot. Some of the stores included Harris & Frank, Silverwoods, Loops Cafeteria, Pet World, Anita Shops, and Raj of India Imports. By 2000, the mall was demolished, and by the next year, the Esplanade Center opened with a Borders, Bed Bath & Beyond, and Circuit City. Few stores remain.

McDaniel's Market, in the Oxnard Super Shopping Center at 2800 Saviers Road, is pictured in October 1954. Other stores to join the Super Shopping Center were W.T. Grants, Thrifty Drug, Wayne's Jewelers, Avila Shoe Repair, and Arcade Flowerland. The McDaniel's Market became several markets over the years: Mayfair Market in March 1961, followed by Fazio's, Valu Plus Warehouse, and most recently El Super.

This corner lot of El Rio was originally known as New Jerusalem and within a short radius included the Simon Cohn Mercantile Store, a couple of saloons and billiard halls run by his brothers David and Leopold Cohn, the Max Gisler residence, and the Santa Clara Chapel. The development of the town of Oxnard took a few of the businesses, and the widening of the freeway took the rest.

Uncle Herb's Pancake House, at 5141 Saviers Road, was started by Herb and Ann Lassiter in 1968. The popular restaurant closed its doors in 2007 after 39 years of business. After the closing of Uncle Herb's Tomas Café, at 622 A Street, brought in Herb's son Ron Lassiter to reintroduce the breakfast menu. Uncle DanMar Restaurant took on the Uncle Herb's site but gave way to the current restaurant, the Best Breakfast Café.

George Omo opened a pool hall on Oxnard Boulevard in 1941 and in 1943 opened Omo's Market at 436 Seaboard Avenue (today Colonia Road). In 1949, he built Omo Market No. 2 at 512 East Date Street. Ten years later, he opened the Omo Motel next door to the market. Today, the 4 Way Meat Market occupies the lot. (Past image, courtesy of Lillian Omo Roman.)

The 20 foot tall Santa was built in Carpenteria in 1950 by Pat and June McKeon to attract customers along the Pacific Coast Highway to a juice bar. Soon, there was a Santa's Kitchen, a Toyland shop, a trading post, and a miniature train. By 2000, the building was declared unsound. In 2003, Mike Barber moved Santa to Oxnard on a lot owned by the Nyland Acres Mutual Water Company. Santa's old perch now shelters the Garden Market.

Hollywood by the Sea was once home to Hollywood movies. As early as 1915, the sand dunes and beaches served as a backdrop to dozens of films before Rudolph Valentino rode the sands of the Oxnard shores in the 1921 movie *The Sheik*. To capitalize on the popularity of the film, developers began buying up the beachfront and selling lots between $290 and $550. Today, the median price for a home is $893,000. The loss of sand dunes led to a loss in interest from Hollywood.

The Port of Hueneme began with Richard Bard. A harbor district was formed in 1937, and a $1.75-million bond was passed. The first ship, the *Margaret Schafer*, brought in lumber from the Northwest. However, with the outbreak of World War II, the port was taken over by the Navy. Commercial operations were reestablished for part of the port after the war. The two images are from 1977 and post-2000, and they show the many changes at the port and the expanding Channel Island Harbor.

The Oxnard Plain hosted several dairies, including the Birra Dairy, pictured around the 1930s; Chase Brothers Dairy; the Williams Dairy at the Scarlett Ranch; and the McGrath Dairy. Desidero Birra was from Switzerland and ran his diary off Rice Road on the Hunter Ranch and leased from the McGrath Estate. Most recently, the Sakioka Family Farms sold its 430-acre parcel to be developed into a business park, including a 576,000-square-foot packing and distributing facility for Arctic Cold and a 75-acre plot for an Amazon fulfillment facility.

Heritage Homes

James Leonard arrived at the goldfields near Marysville from Ireland. He worked his way towards Berkeley and purchased land he later donated for the university. Leonard arrived in the southern portion of the state in 1868 and purchased 1,000 acres next to the Santa Clara River at the corner of present-day Gonzales Road and Victoria Avenue. He imported lumber to build the first wood-frame structure on the Oxnard Plain, pictured here in 2005.

Heritage Square is located between Seventh and Eighth Streets and A and B Streets. Dedication took place on October 24, 1991, with speeches from Mayor Nao Takasugi and the principal contractor, Patrick McCarthy. Many of the descendants of the repurposed homes were also present, including the Connolly, Fry, Gordon, Hartveld, Laurent, Maulhardt, McGrath, Perkins, Petit, Petre, Pfeiler, Snively-Ruggles, and Scarlett families. The 3,000-gallon water tank and tower from the Pfeiler ranch were also relocated. Today, the barn and farming remain. (Past image, courtesy of Gary Blum.)

Justin and Frances Kaufman Petit's home was at 1900 East Wooley Road. In 1896, Petit hired architect Franklin Ward and contractor Herman Anlauf to build a 7,100-square-foot home with 13 rooms, including 7 bedrooms. The Queen Anne Victorian–style house was the first farmhouse in the county to be lighted by electricity. The restoration meticulously retained many of its original features, including the spindle and spool work on the verandas, curved glass for the corner windows, and two chimneys. (Past image, courtesy of Gary Blum; present image, courtesy of Aurelio Ocampo Jr.)

The 4,200-square-foot Perkins house was at 464 Pleasant Valley Road and was built in 1887 by Jens Rasmussen in the stick style of Queen Anne architecture. It was built for David Todd Perkins and later sold to Charles Rowe and his wife, Mary Donlon Rowe, in 1904. They held many large barbecues, including one for the New York Giants in 1913. The last family to reside at the house was the Clay Claberg family, who purchased the 23 acres in 1919. The current owner is Al Barkey. (Present image, courtesy of Aurelio Ocampo Jr.)

The oldest of the group of buildings at Heritage Square is the Italianate-style home of Louis and Caroline Pfeiler. Built in 1877, it was located at 1980 Rice Road. The front and side porches featured round arched and filigreed wood designs. Pfeiler arrived in the county from Austria in 1872. He married one of the Kaufman daughters, who married men from the Borchard, Hartman, and Petit families as well, creating many distant cousins to this day. (Present image, courtesy of Aurelio Ocampo Jr.)

The Christian Church was built in 1906 at the corner of Second and D Streets. It is built in the Carpenter Gothic style and features lancet arched stained-glass windows and decorative gable trim. The building also featured a Mission Revival–style parapet at one end. In 1915, the First Church of Christ purchased the building from a donation by local farmer Thomas Rice. (Present image, courtesy of Aurelio Ocampo Jr.)

The Scarlett House was built by J.W. Parrish in 1903 at 211 South C Street. The building features the transitional period between the Queen Anne and Colonial Revival architectural styles, with a primary emphasis on the Colonial style. A main feature is the front window, which is divided into three parts separated by Classical-style columns and capped with a raised floral motif. The home now serves as a wine tasting venue, Rancho Ventavo Cellars. (Past image, courtesy of Gary Blum; present image, courtesy of Aurelio Ocampo Jr.)

C. St. Looking South, Oxnard, Cal.

The Fry-Puntenney House at Heritage Square was built in 1903 for Abraham Fry and his wife, Elizabeth, at 210 South C Street. Later, Harriett Puntenney gave music lessons from the house for years. The house features a Queen Anne–style tower and Classical-style porch columns connected by a turned balustrade and decorative paired brackets under the bellcast porch roof. The house also features a Palladian window. (Present image, courtesy of Aurelio Ocampo Jr.)

The Colonial Revival–style Laurent-McGrath House was built in 1901 for Martin Laurent and his wife, Annette Petit Laurent, at 403 South C Street. Later, the house was purchased by John McGrath, who by the 1960s moved it to Wooley Road. The house features Classical porch columns, diamond-paned shaped windows, and decorative eave brackets. The first floor features wide clapboard siding, while the second floor has narrow siding. The house is home to restaurant and speakeasy La Dolce Vita 1901. (Past image, courtesy of Gary Blum; present image, courtesy of Aurelio Ocampo Jr.)

The Colonial Revival Patterson-Hartveld House was built in 1905 for Patterson Ranch manager John Roupe. After the ranch was subdivided in 1917, Peter Hartveld bought the house, and family members lived in it until 1987. The house has Classical-style porch columns and Corinthian capitals plus three-part Palladian windows under the side gable. It is clad in narrow clapboard siding with diamond-shaped shingles under the side gables. (Past image, courtesy of Gary Blum; present image, courtesy of Aurelio Ocampo Jr.)

The Snively-Ruggles House was built in 1885 for W.R. Snively. His daughter Bernice married Earl Ruggles and owned the property until the 1980s. The two-story farmhouse design was influenced by the Colonial Revival style and features a long porch with a balcony and balustrade above. Unfortunately, the original house burned down just before it was to be moved, so the house was replicated. (Past image, courtesy of Gary Blum; present image, courtesy of Aurelio Ocampo Jr.)

The Gordon House was built in 1910 for Alonzo and Sara Wood Gordon at 2151 Wooley Road. He was an ordained Baptist minister and organized the Baptist church in Springville, a small rural community between Oxnard and Camarillo. The Craftsman-style home features a steep-pitched gable roof that sweeps down over the porch. Under the eaves are triangular braced supports, notched rafters, and exposed beams. The porch features brickwork and massive capped square columns. (Past image, courtesy of Gary Blum; present image, courtesy of Aurelio Ocampo Jr.)

The Archie Connolly House was built in 1912 on the Connolly Ranch off Gonzales Road west of Ventura Road. Connolly married Eliza Cloyne, from another early Irish family in Oxnard. The house was built in the Craftsman and Prairie styles, featuring a two-story recessed porch and balcony and a two-story slanted bay window. The house was designed by prominent Los Angeles–based architect Albert C Martin. (Past image, courtesy of Gary Blum; present image, courtesy of Aurelio Ocampo Jr.)

The Maulhardt Winery also served as a storehouse for Gottfried and Sophie Maulhardt at their ranch at 201 Rose Avenue. The two-story brick building included a portion of the first floor constructed below ground level. Here, the wine was stored for the Santa Clara Chapel. Due to the fragile nature of the original 1876 building, the winery was replicated by Patrick McCarthy's crew. The original winery remains intact and reinforced at the Oxnard Historic Farm Park. (Present image, courtesy of Aurelio Ocampo Jr.)

The original cottage home was built by Gottfried Maulhardt in the 1870s. His widow, Sophie, sold the ranch to Louis Pfeiler in 1904, and he gave the ranch to his son Albert Pfeiler and wife Lydia as a wedding present. Albert built several additions to accommodate his growing family. Albert's son Robert was the last to farm the ranch up until 2002. The house is now part of the Oxnard Historic Farm Park at 1251 Gottfried Place and designated Ventura County Historical Landmark No. 165.

ROW OF BUNGALOWS, OXNARD, VENTURA CO., CAL.

The houses on F and G Streets between Palm and Fifth Streets are part of the Henry T. Oxnard Historic District. The district includes 139 buildings constructed between 1909 and 1941, and they represent a variety of styles, including Craftsman, Spanish Colonial, and Period Revival bungalow. The district was placed in the National Register in 1999 and also serves at the city's Christmas Tree Lane during the holidays.

The Hill residence, at 737 West Sixth Street, was part of the ranch owned by Jack Hill and his wife, Aranetta Rice Hill. They sold 31 acres to Henry Oxnard that became the townsite, thus making this the first home in the town of Oxnard. Aranetta Hill fought against incorporation in 1903 and was one of 13 who voted against cityhood versus the 283 who were in favor. The residence later served as home to the Monday Club and then became the Korean Presbyterian Church.

Consistent with our mission to preserve history on a local level, this book was printed in South Carolina on American-made paper and manufactured entirely in the United States. Products carrying the accredited Forest Stewardship Council (FSC) label are printed on 100 percent FSC-certified paper.